Herbert Durand

Hot Springs, Arkansas

Its hotels, baths, resorts and beautiful scenery

Herbert Durand

Hot Springs, Arkansas
Its hotels, baths, resorts and beautiful scenery

ISBN/EAN: 9783337419509

Printed in Europe, USA, Canada, Australia, Japan

Cover: Foto ©Andreas Hilbeck / pixelio.de

More available books at **www.hansebooks.com**

Hot Springs

Arkansas

Its
Hotels, Baths, Resorts
and
Beautiful Scenery

ILLUSTRATED

PRESS OF
THE WOODWARD & TIERNAN PRINTING CO., ST. LOUIS

How to Get to Hot Springs, Arkansas

AND WHO TO SEE OR ADDRESS FOR FULL INFORMATION IN REGARD TO RATES, ETC

From Omaha, Lincoln, St. Joseph, Atchison, Leavenworth and Kansas City.
Mo. Pac. R'y to St. Louis, Iron Mountain Route to Hot Springs without change; or Mo. Pac. R'y to Coffeyville, Kan., and Wagoner Route, via Ft. Smith and Little Rock, to Hot Springs.

- C. E. STYLES, Passenger and Ticket Agent...ATCHISON, KAN.
- J. N. JOERGER, Pass'r and Tkt Ag't LEAVENWORTH, KAN.
- F. P. WADE, Passenger and Ticket Agent...ST. JOSEPH, MO.
- R. P. R. MILLAR, General Agent...... ...... LINCOLN, NEB.
- T. F. GODFREY, Passenger and Ticket Agent..OMAHA, NEB.
- G. E. DORRINGTON, Traveling Pass'r Agent..OMAHA, NEB.
- J. H. LYON, Western Passenger Agent,
  800 Main Street, KANSAS CITY, MO.
- E. S. JEWETT, Passenger and Ticket Agent,
  800 Main Street, KANSAS CITY, MO.
- BENTON QUICK, Passenger and Assistant Ticket Agent,
  1048 Union Avenue, KANSAS CITY, MO.

From St. Louis, Memphis, Little Rock, Ft. Smith, Helena, and Texarkana.
Iron Mountain Route, through to Hot Springs.

- W. H. MORTON, Passenger Ag't,Union Depot, ST. LOUIS, MO.
- M. GRIFFIN, City Pass. Agt, cor Bdwy & Olive, ST. LOUIS, MO.
- H. D. WILSON, Passenger and Ticket Ag't. MEMPHIS, TENN.
- H. F. BERKLEY, Pass'r and Ticket Ag't..LITTLE ROCK, ARK

From Salt Lake, Denver, Colorado Springs, Pueblo, Nevada, Carthage, Wichita, and All Points in Southwestern Missouri and Southern Kansas.
Mo. Pac. R'y to Coffeyville, and Wagoner Route, via Ft. Smith and Little Rock, to Hot Springs.

- C. A. TRIPP, General Western Freight and Passenger Agent,
  30-2 Larimer St., DENVER, COLO
- F. E. HOFFMAN, Traveling Passenger Agent, DENVER, COLO.
- WM. HOGG, Ticket Agent.................PUEBLO, COLO.
- S. V. DERRAH, Commercial Passenger Agent,
  21 Morlan Block, SALT LAKE CITY, UTAH.
- E. E. BLECKLEY, Passenger and Ticket Agent,
  224 E. Douglas Avenue, WICHITA, KAN.

From Chicago, Milwaukee, St. Paul, Minneapolis, Des Moines, Quincy and the Northwest.
Any direct line to St. Louis, and the Iron Mountain Route, St. Louis to Hot Springs, without change.

- JNO. E. ENNIS, District Passenger and Land Agent..................................220 S. Clark Street, CHICAGO, ILL.
- H. D. ARMSTRONG, Traveling Passenger Agent..JACKSON, MICH.

From New York, Boston, Philadelphia, Washington, Pittsburgh, Buffalo, Cleveland, Toledo, Detroit, Cincinnati, and Indianapolis.
Any direct line to St. Louis, and the Iron Mountain Route, St. Louis to Hot Springs, without change.

- W. E. HOYT, Gen'l East'n Pass'r Agt., 391, Brdwy, NEW YORK.
- J. P. McCANN, East'n Trav. Agt., 391 Broadway, NEW YORK.
- G. K. DELAHANTY, New England Passenger Agent,
  290 Washington St., BOSTON, MASS.
- N. K. WARWICK, Dis't.Pass.Agt.,Ell Vine St., CINCINNATI, O.
- S. H. THOMPSON, Central Passenger Agent,
  1129 Liberty St., PITTSBURGH, PA
- COKE ALEXANDER, District Passenger Agent,
  7 Jackson Place, INDIANAPOLIS, IND.

From Richmond, Savannah, Atlanta, Charleston, Chattanooga, Nashville, Knoxville, Birmingham and the Southeast.
Any direct line to Memphis, and the Iron Mountain Route, Memphis to Hot Springs.

- A. A. GALLAGHER, Southern Passenger Agent..105 Read House, CHATTANOOGA, TENN.

From Louisville, Evansville, Lexington and Frankfort.
Any direct line to St. Louis or Memphis, and the Iron Mountain Route to Hot Springs, without change.

- BISSELL WILSON, South'n Trav. Agt., 336 W. Main St., LOUISVILLE, KY.
- J. W. MASON, Passenger Agent.....CAIRO, ILL.

From Galveston, Houston, Velasco, Palestine, City of Mexico, Laredo, San Antonio, Austin and San Marcos.
I. & G. N. R. R. and T. & P. R'y to Texarkana, and the Iron Mountain Route, Texarkana to Hot Springs.

- J. C. LEWIS, Travelling Passenger Agent..AUSTIN, TEXAS.

From San Francisco, Los Angeles and All Points in California, Arizona and New Mexico.
So. Pac. to El Paso, T. & P. to Texarkana, and the I. M. Route, Texarkana to Hot Springs.

- L. M. FLETCHER, G. P. C. Freight and Passenger Agent.................................660 Market Street, SAN FRANCISCO, CAL.

From Colorado City, Abilene, Ft. Worth, Sherman, Denison, Paris, Dallas, Terrell, Marshall and Jefferson.
Texas & Pacific to Texarkana, and the Iron Mountain Route to Hot Springs.

From New Orleans, Baton Rouge, Shreveport, Alexandria, and All Points in Northern Louisiana.
Texas & Pacific Railway to Texarkana, and the Iron Mountain Route to Hot Springs.

- J. C. LEWIS, Travelling Passenger Agent, AUSTIN, TEX., or any Ticket Agent Texas & Pacific Railway.

From Mobile, Meridian, Jackson, Vicksburg, and All Points in Mississippi and Alabama.
Any direct Line to Arkansas City or Memphis, and the Iron Mountain Route to Hot Springs.

- A. A. GALLAGHER, Southern Passenger Agent,
  105 Read House, CHATTANOOGA, TENN.
- H. D. WILSON, Pass'r and Ticket Ag't.....MEMPHIS, TENN.

Object of ye Picture Booke.

Yᵉ Suffering publick. Sobeit yᵗ yᵉ ills be of yᵉ flesh, or mayhap of yᵉ minde, will finde depicted in yᵉ pages of this booke a localitie wherein all yᵉ foule disorders, distempers, humours, vapours, aches, tumours, paines, twinges, ennuyés & such like afflictions may be dispelled from yᵉ bodie – & yᵗ right Speedilie··

Yᵉ Seeker after frivolities, such as picturesque scenerie, romantick ramblings on horseback or on shanks, his mare, hunting of yᵉ wilde beaste, or birde, or fish, dancing, coquetrie & all such social fripperie, will finde here opportunities & facilities for indulging yᵉ lightsome moodes & fancies to yᵉ Quean's taste – & yᵗ right merrilie.

& bothe of yᵉ aforesaid sortes & conditions of yᵉ publick will finde in yᵉ inns of yᵉ Hot Springs such genial hostelrie, such commodious lodgings and such goode cheere for yᵉ inner man, yᵗ yᵉ guests are feign to indulge in yᵉ moste extravagant lauds & flatteries – & yᵗ right heartilie.

Note. – Yᵉ artiste having in his inscrutable maieste seen fit to inscribe yᵉ title of yᵉ booke in yᵉ ancient style, yᵉ compiler hath found it consistent to write yᵉ preface accordinglie. This concession to the congruities accomplished, he now proposes to proceed in plaine United States – & yᵗ right willinglie.

CONTENTS.

	PAGE.
How to get to Hot Springs	2
Object of ye Picture Booke	3
The Trip to Hot Springs	7
The Hotels of Hot Springs	15
The Hotel Eastman	16
The Park Hotel	24
The New Arlington Hotel	32
Other Hotel Accommodations—Hotel Hay—Hotel Worrell—Pullman—Avenue Hotel—Plateau—New Waverly—Sumpter House—Hotel Josephine—Albion—Magnolia Villa—Taylor's—Haynes Villa	37–42
The Springs and Their Properties	43
The Bath Houses	45
The City of Hot Springs	47
Happy Hollow	60
The Man on Horseback	61
Carriage Drives	65
Potash Sulphur Springs	66
Gillen's White Sulphur Spring	68
Mountain Valley Springs	75
The Ouachita River and its Inhabitants	74
Hot Springs in Legend and History	75
Statistical Table	80

ILLUSTRATIONS.

	PAGE.
Graphic Scenes from Hot Springs of the past	5
The Old Way and the New	6
Malvern Station	8
First Glimpse of the Ouachita	8
Wood Hauling with Oxen	9
Cove Creek	10
Lawrence Station	11
Scenery on the Hot Springs R. R.	12
Glimpses of the Gulpha, Hot Springs R. R.	13
Arrival of Train at Hot Springs	14
The Hotel Eastman and Surroundings	16, 17
Children's Donkey Party	18
Grand Promenade, Hotel Eastman	19
Rotunda and Ball Room, Hotel Eastman	20
Grand Parlor, Hotel Eastman	21
Dining Room, Hotel Eastman	22
Views from Eastman Tower	23
Park Hotel and Grounds	25
Observatory, Park Hotel	26
Grand Parlor and Staircase, Park Hotel	27
Grand Dining Room, Park Hotel	28
Dancing Hall and Pavilion, Park Hotel	29
Scenery near Park Hotel	30
Park Hotel Bath Rooms	31
The Old Arlington	32
The New Arlington	34
Views from the Cupola New Arlington Hotel	36
Hotel Hay	38
The Pullman	38
Hotel Worrell	38
Avenue Hotel	39
The Plateau	39
Waverly Hotel	40
Sumpter House	40
Magnolia Villa	41
The Albion	41
Taylor's, Park Ave.	41
Haynes Villa	41
Hotel Josephine	41
Hot Springs Mountain	43
Glimpse of Bath House Row, Gov't Reservation	43
View from Prospect ave.	44
Bath Houses	46
Bird's-eye View of Hot Springs, 1891	47
Hot Springs Creek	48
Grand Opera House and Post Office	49
Hot Springs Business Blocks	50
Hot Springs Water Works	52
Hot Springs Churches	53
Government Buildings, Hot Springs	54
Hot Springs Residences	55–59
West Mountain from Happy Hollow	60
Happy Hollow from West Mountain	60
Scenes in Happy Hollow	61
Sketches on Hot Springs Mountain	62
Donkey Station, etc., Happy Hollow	63
Chalybeate Springs	64
Boulevard Drive, Hot Springs Creek	65
Potash-Sulphur Springs	67
White Sulphur Springs	69
Hell's Half Acre	70
Scenes on the Gulpha	71
Mountain Valley Springs	72
Scenes on the Ouachita	75
Bull Bayou, Mountain Stream	76
Wheeler's Ford and Ferry	78

Graphic Scenes
— from —
Hot Springs
of the
Past.

Hot Springs as it was — Drawn by Prof. D. D. Owen in 1859.

MALVERN STATION — Junction of the Iron Mountain Route with the Hot Springs R. R.

The Trip to Hot Springs.

THERE is no question that the trip to Hot Springs a decade ago was attended with a few inconveniences, especially after the traveler left Malvern and the railroad. The huge lumbering stage coach was cramped, stuffy and shaky; the driver uncouth and profusely profane; the streams to be forded frequent, frolicsome and deep; the mud likewise, except where the way was stony—and there it was hilariously frolicsome, you can be sure—the wild and woolly road agents, with regulation black masks and six-shooters, were ubiquitous, aggressive and acquisitive, and the journey such a twenty-five rugged, jostling, weary miles, that it delivered the venturesome passengers at the Springs in a condition of physical, mental and frequently of financial collapse.

If the gentle reader will turn back to page 5, he will find there a graphic portrayal of the wonderful transformation in the methods of transportation through this region. The stage coach has given way to the Pullman palace car; the boisterous driver to the gentlemanly conductor; the frolicsome ford to the Howe truss steel bridge; the mud and the rocks to steel rails and stone ballast, and the black-masked road agent to the black-skinned sleeping car porter.

How thoroughly enjoyable is this journey through Arkansas nowadays! What luxury lurks in a Pullman car—that drawing-room of dreams by night, and parlor of panoramas by day! Every window is a frame for kaleidoscopic landscapes—here a cotton, there a corn field, livened by their dusky laborers; then a glimpse of some broad, majestic river, and again stretches of dense and sombre forests, void of the presence of man, fresh and fragrant in spring, cool and restful in summer, flaming and magnificent in autumn, and relieved of dreariness, even in winter, by the glorious greens of the holly, pine and cedar.

The train stops at Little Rock, the State capital, but the passengers are more interested in the sumptuous breakfast served at that desideratum of travelers, a really excellent railway eating house, which actually exists here. The meats, milk, eggs and butter are "out of sight," in a slangy, and soon in a literal sense, and away we go for the Springs. Malvern, the junction point of the Iron Mountain and Hot Springs railroads, is, now that a solid train is run between St. Louis and the Springs, reduced to the menial position of a way station; so, without change of cars or transfer of baggage, the last stage of the trip is begun.

The character of the landscape now begins to change; the country becomes hilly; the locomotive strikes heavy grades, and snorts and puffs and groans its way to the summits. Occasionally a limpid, hurrying, cascade-bedecked streamlet jumps from the hillside, bustles through a culvert, and scampers away into the thicket like a mischievous urchin. Suddenly some one exclaims, "There's the Ouachita!" (Say "Washytaw," please, not "Oochytaw," as if you had sat on a bent pin), and everyone rushes to one side of the car to catch a fleeting glimpse of a noble stream, as it sweeps majestically in a half circle around, and laving the feet of a lofty cliff We will go a fishing in the Ouachita later on, if you like, and we will lie about our catch, also, I doubt not. A little further on a country road meanders in and out of the pines—the road of the stage coach, the fords, the mud, and the robbers, perhaps — and one sees, toiling patiently along, a yoke of sluggish oxen, slowly dragging their rude lumber-laden cart, and urged on by the "Gees" and "Haws" of the jeans-clothed native, and the prick of his galling goad.

FIRST GLIMPSE OF THE OUACHITA

The train darts by a little station; one catches the words "Cove Creek," in black letters. We don't stop now, for the reason that the locomotives of the Hot Springs Railroad have risen to the dignity of coal burners. A couple of years ago they burned wood, and we stopped, and the train crew and passengers—every mother's son of them that could walk and work their arms—used to get out and help "wood up." The old visitor to Hot Springs misses this episode; he also misses the fragrance of the penetrating, pungent, tarry smoke that was wont to belch out with such billowy and blinding bountifulness—and rejoices thereat. I find it stated in a publication descriptive of Hot Springs, that "Cove Creek abounds in trout," and desire to inform the enthusiastic angler that "trout" is, in Arkansas, a generic term, applied indiscriminately to any kind of fish caught in small streams, but in particular to the black bass. I doubt if there are any brook trout in Arkansas waters, except at Mammoth Springs. There are certainly none in Cove Creek. But, to return to our mutton—the railroad train. We will now proceed to labor up further steep grades, shoot around curves so sharp that the rear brakeman can almost shake hands with the engineer, rumble over bridges, until, presto! we find ourselves surrounded by mountains and realize that we are, at last, in the "Heart of the Ozarks" and nearing our destination.

A short halt is made at Lawrence, some seven miles from Hot Springs, where a handsome station has been erected for the accommodation of visitors to Potash Sulphur Springs—a resort one mile distant, of great virtue, which will be described further on. As we leave Lawrence, we catch frequent glimpses of the picturesque Gulpha, a charming rivulet, which, like its sister streamlets, shimmers and sparkles through many a mile of wooded glen, on its babbling, bounding course to the Ouachita. And now let us devote the few minutes remaining, before the brakeman cries, "All out for Hot Springs," to a brief history of the wonderful little railroad over which we have journeyed for the past hour.

One afternoon in February, 1874, three gentlemen alighted at Malvern from an Iron Mountain train on their first visit to Hot Springs. They were "Diamond Jo," Reynolds, Col. L. D. Richardson and Capt. William Fleming. A complete canvass of the town failed to discover anyone who would agree to drive them to the Springs that day for any amount of money, so they were compelled to remain over night. Next morning a Jehu was found, who, after much bickering, agreed to convey the parties to their destination for six dollars each, payable in advance. Protesting, but necessarily agreeing

Wood Hauling with Oxen.

LAWRENCE STATION, Hot Springs R.R.

to this exorbitant charge, the trio mounted the rickety vehicle provided them and set forth. Six miles from Malvern, and twenty from Hot Springs the wagon broke down, and the driver announced that he would be compelled to return to Malvern for another one. Disgusted with their ill-fortune, and already weary of the horrible jolting and snail-like progress of those first six miles, it was proposed and agreed that they should finish the journey on foot, which they proceeded to do, reaching the Springs late that night, but, nevertheless, ahead of the dilatory driver, who, having gone back to Malvern and secured another vehicle, followed them, an hour or so later, with their baggage. It was during this long, tiresome walk that the Hot Springs Railroad project was conceived. "By George, Jo.," said Col. Richardson. "there ought to be some better way of getting over here than walking, or riding in one of these infernal cow carts."

"That's so," answered Diamond Jo.; "say, Rich, let's build a railroad. If people will come from all over the world to visit these Springs, and undergo such experiences as we have, there must be something in them, and a railroad would not only develop the place, but, I believe, would pay handsomely."

The conversation, thus begun, was continued as they plodded along, and, after reaching the Springs and before going to bed that night, it was agreed that Col. Richardson should take the earliest opportunity to confer with President Allen, of the Iron Mountain Railroad, and obtain from him such information and co-operation as would enable them to get the project under way. This conference took place in St. Louis a short time afterwards. It was learned that the Iron Mountain held a charter for a standard gauge road, but, on account of the stringency of money at that time, did not feel warranted in incurring the expense necessary for construction. As Diamond Jo. was enthusiastic over the scheme, however, and, as financial panics had no effect on

his pocket-book, the latter difficulty was quickly overcome, and, a charter having been obtained from the Legislature, with the assistance of President Allen, ground was soon broken, and the autumn of the same year in which the three pedestrians plodded their weary way, saw a narrow gauge railroad in full operation from Malvern to Lawrence. The original fare was ten cents per mile, or $2.50 from Malvern to the Springs, the distance then being twenty-five miles. This fare was subsequently reduced to $2.00, then to $1.60, and, finally, to $1.10, the distance having been reduced to twenty-two miles by straightening curves. In October, 1889, the road was changed from narrow to standard gauge, and, in January, 1890, the first Pullman sleeping car ran through from St. Louis to Hot Springs. This service, with such additions and improvements as experience and increasing patronage have suggested, has continued to the present day. Col. L. D. Richardson, one of the original projectors, some years ago took the position of general manager, and later, upon the death of Mr. Reynolds, became also president of the road, which position he holds to-day. Under his administration, a large amount of money has been expended in betterments, the line has been relaid with heavy steel rail, steep grades modified, new bridges built, and other improvements made, until to-day the property is in first-class condition throughout. Branch railroads, as a general rule, are but sorry affairs, with poor roadbeds, decrepit rails and rattle-trap rolling stock, to say nothing of the usual freight train time, irregular running and utter lack of accommodation. The Hot Springs Railroad, however, is a bright and shining exception, and to the traveler there is nothing about the track, cars or the time, to indicate that he is not on a main line of some great system. Three passenger trains are run each way daily between Hot Springs and Malvern, connecting with all Iron Mountain trains and making the Springs perfectly accessible, without delay, from every part of the country. At the Hot Springs station, coupon tickets can be purchased and baggage checked through to any point in the United States, every facility for the convenience of travelers being provided. The Pacific Express Company operates over this line and has an office at the Springs.

Well, there goes the whistle, and the passengers are collecting

their grips, wraps and umbrellas. The train stops at a tasty little brick station, backed by a handsome park shaded with lofty forest trees, and we disembark. The entire population is here to meet us—for it is one of the features of the daily routine to go to the trains to welcome the coming and speed the parting guests. And what a cosmopolitan crowd it is! English, French, Germans, Spaniards, Italians, Greeks, Turks, Russians, Arabs, Africans—natives mostly—Indians, Chinese—they are all here; and of our own Americans, every State in the Union is represented. Yet it is a good-natured, jolly, jostling crowd withal, with a general air of *bonhomie* that is contagious and makes one feel quite at home at once.

As one drives to his hotel from the station, he is vividly impressed with the remarkable contrasts visible on every side. Strong and vigorous men pass, with manly stride, their antipodes in invalid chairs or on crutches; the meek oxen gaze in silent wonderment at the spanking team of thoroughbreds which prances by; the stylishly dressed New Yorker or Londoner strolls along, side by side with the Ozark farmer in his ragged suit of gray jeans; the very buildings share the general antithesis; handsome four-story bricks look down on crude one-story wooden shanties, and colossal hotels overshadow ramshackle lodging houses. These are the features that make Hot Springs so unique, picturesque and interesting. A recent writer concludes his article on the Springs as follows:

"There is not an hour in the twenty-four that one cannot be entertained here to the full limit of his tastes, from a church fair to a cake walk, a milkmaid's convention to a Y. M. C. A. lecture, or a good sermon to a jack-pot or a prize fight. Hot Springs stands *sui generis.*"

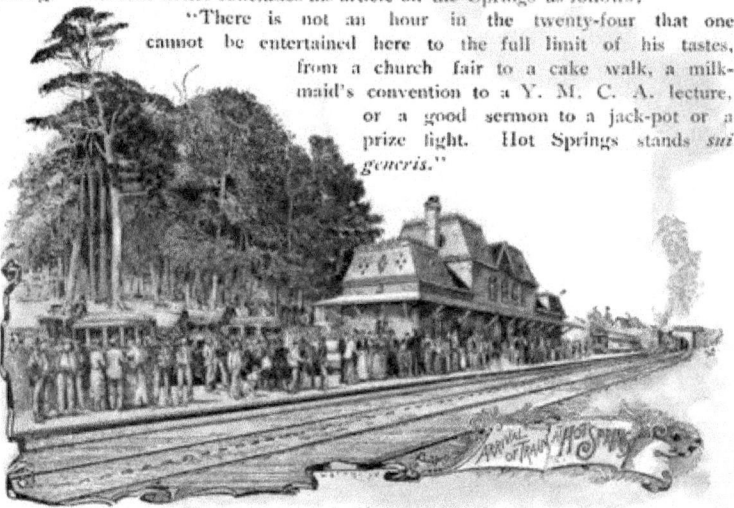

The Hotels of Hot Springs.

IN selecting a hotel there are always several things to be considered before reaching a decision. One may be rich, and wish to surround himself with every luxury; or poor, but still desirous of all possible comfort; an invalid, in search of perfect quiet and careful attendance; or in robust health, and on pleasure bent; of strong social tendencies and looking for the companionship of Fashion's devotees; or of a retiring nature, seeking only for seclusion and rest. The patrons of Hot Springs number many of each of these classes, and many more whose tastes and requirements diverge still more. Nevertheless, this Arkansas resort is fully equal to the emergency. Its five hundred hotels and boarding houses are of all grades and suitable for all sorts and conditions of men. There is no place of the kind in the country—perhaps not in the world—where every one, no matter what his social, financial or physical condition may be, can find an abiding place perfectly adapted to his needs, so readily as at Hot Springs.

To the wealthy seeker, after either health or recreation, he who is able and willing to pay for the best of everything, the Eastman, the Park and the New Arlington, offer accommodations not surpassed by any hotel in America. He can have rooms *en suite*, sumptuously furnished, with private bath; he is assured of an unexceptionable cuisine; he can surround himself with luxuries *ad libitum*. Those who do not care for such ultra-expensive lodgings, but who desire, nevertheless, the best possible living, will find in the same hostelries, accommodations exactly suiting their tastes. The large number of hotels of the middle class, excellent in appointment and management, but not on so lavish a scale as the houses named, find their patronage largely among those who visit the Springs in search of health and do not feel warranted in going to any extraordinary expense. On this class also, the numerous smaller hotels, villas and private boarding houses draw largely, and the wide range of price, comfort and location is such that, as before intimated, any one can satisfy himself according to his resources.

With a view of giving the intending visitor the best possible information, and thereby aiding him in determining the all-important question of how and where he shall live while at the Springs, and the approximate expense, the following pages are devoted to detailed descriptions of the various leading hotels, and to general details as to the smaller hotels and boarding houses.

The Hotel Eastman.

THE building of the Eastman was the beginning of the development of Hot Springs into an all-the-year-round fashionable resort for rest and recreation. With its completion, the army of invalids constantly marching to this modern Mecca of health, found its ranks reinforced by robust representatives of the wealth and culture of the nation. There were to be seen gay groups of pedestrians and equestrians by day, and there were sounds of revelry by night.

The hotel sprang up like another Aladdin's palace, but eight months elapsing from the beginning of the structure, in May, 1889, until it was ready for occupancy. It is an imposing five-story building, of colossal dimensions, covering several acres

of ground, and crowned with lofty towers and observatories which overlook the Ouachita Valley and the peaks of the Ozarks for miles and miles. It is constructed on two sides of a quadrangular park, decorated with trees, flowers and fountains, forming a delightful approach. The hotel contains five hundred and

twenty guest rooms, all large, well lighted and elegantly furnished and appointed. Each room may be considered a front one, as there are none but command delightful views of valley, mountain, stream or woodland. The main halls, twelve feet wide, extend through the center of the entire building, each forming a grand promenade six hundred and seventy-five feet long.

Children's **Donkey** Party.

No fire is ever lighted in the house except in the magnificent fire-places in the parlors and office, and in the kitchen, which is positively fire-proof. The building is heated throughout by steam and lighted by electricity, both the incandescent and the arc systems being used. There are few resort hotels in the country that have made such bounteous provision for the comfort and convenience of their guests as the Eastman. Besides the grand rotunda, 52 by 70 feet, and the grand parlor, there are a ladies' parlor, ladies' reading room, gentlemen's parlor and gentlemen's reading room, ladies' billiard room, card and reception rooms, and writing rooms galore. All are most sumptuously furnished and fitted with every necessary accessory.

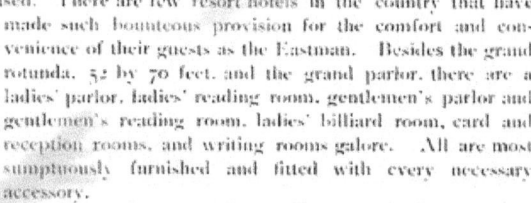

The bath house and its appointments are simply superb. It is located east of the hotel and across Cottage avenue, but a corridor built across this avenue connects the two buildings.

the corridor being an extension of the second-story hall. Both corridor and bath house are heated by steam, insuring an equable temperature to the bathers en route to and from their rooms. There are eight parlors and forty bath rooms, the latter constructed entirely of brass and marble, and the bath tubs lined with the most expensive Roman porcelain. The hot water is brought from the government reservoir far above, on Hot Springs Mountain.

The spacious park in front of the Eastman, above referred to, forms a great romping ground for the children, who can be seen at all hours of the day busy at their games, rolling and tumbling on the grass, riding the obstinately slow but persistent burros, laughing and screaming with delight, while their more sedate elders

Grand Promenade, Hotel Eastman

look on complacently as they promenade along the broad verandas or rest quietly in the huge but cosy rocking chairs.

The observatory tower is a popular addition to the Eastman, rising to an elevation of nearly two hundred feet, and revealing to the guest who scales its dizzy height a magnificent cyclorama of mountain and vale and forest streams, which well repays the exertion of the ascent.

The daily routine at the Eastman is literally one continual round of pleasure. The hotel is blessed with a superb orchestra, which discourses sweet music morning,

Rotunda and Ball Room.

afternoon and night in the grand rotunda. So popular is this orchestra that it is a regular fad for the guests of other hotels to organize parties to visit the Eastman and listen to the concerts. At nine o'clock each night the music adjourns to the grand ball room and furnishes rythmic inspiration for an assemblage of merry dancers. A german is given at least once a week, and square and round dances are in vogue on other nights. This, with card parties, theater parties, exploring trips among the mountains, horseback excursions in the country, and the other numerous amusements always suggested at pleasure resorts, make the life of the Eastman guests a truly happy one during their sojourn at the Springs.

Dining Room.

The season at the Eastman usually runs from early in January until June 1st, announcements as to the exact time of opening and closing being made in ample time each year. The present season of 1892 is the third, and the large patronage, even exceeding the capacity of the hotel, shows its great popularity. Its guests register from all quarters of the globe.

The attendance at the Eastman is unexceptional. From the manager down to the bell boys, the sole and constant aim of each attache appears to be to insure the comfort and pleasure of the patrons—no easy task, when one thinks that they have a population that would make a small city of itself, to look after and care for, to lodge, amuse and feed—and that reminds me, that any description of the Eastman omitting reference to the grand dining room and its splendid service would be playing Hamlet with Hamlet left out.

The first impression, as one enters this stately and extensive hall, is one of mingled wonder, bewilderment and admiration; wonder at its colossal dimensions, bewilderment at the gorgeous spectacle afforded by the myriad lights and the gaily dressed multitude, and admiration of the beautiful and harmonious decorations. A thousand people may be seated here and served speedily and satisfactorily by the army of thoroughly trained waiters. The menu is not surpassed by any hotel in America. Thanks to the perfection of the refrigerator car system, the choicest meats, game and delicacies from all parts of the world can be and are served to guests here, as fresh and delicious as if on their native heath. There is never lack of variety, either in the food or its preparation, which shows the chefs to be masters of their art—the true art preservative.

THE PARK HOTEL.

YOU have noticed that handsome five-story brick, over yonder on Malvern avenue? Well, that is the Park Hotel, and we are about to investigate it. Though but two blocks from the railway station, these two blocks are in the opposite direction from the noise and bustle of the business houses on Central avenue, and we find ourselves on a quiet, shaded street, not unlike that of a peaceful country village, but for the street cars which run to and from the main part of the city. We are naturally impressed with the architectural beauty of the hotel, with its pleasing promise of quiet comfort, and with its charming surroundings. Located on an eminence in a natural park of some ten acres, with grassy, flower-bedecked lawns, and lofty trees, it commands an unobstructed view on every side, of the picturesque Ouachita Valley and the encircling arms of its mountain lover—the Ozark range. This bright sunny afternoon the broad verandas are gay with guests, some promenading slowly to and fro; others engaged in a go-as-you-please contest—six times around and back to the mile; still others in easy chairs, enjoying the warmth and geniality around them. A merry party of young people come out, mount their waiting horses and dash away for an excursion among the mountains. Inside, the same air of cheerfulness prevails. The rotunda is superb in its ornamentation, brilliantly lighted from all sides, and enlivened by the presence and conversation of a hundred people, who gather in jolly, chattering groups, or loll lazily in the huge rocking chairs, behind a paper or the latest novel, while, at short intervals, a fine orchestra drowns, with its melody, the euphonious hum of busy voices.

Under the ciceronage of an affable young man, with silvery hair, a silvery tongue and a bunch of brassy keys, we are conducted through the hotel. We are told, as we start, that there are two hundred and seventy-five guest rooms, all front rooms, and all equally well furnished and fitted. We pass through a short hallway, near at hand, and are shown into the bath house—in a separate building, but so near that it is but fifty steps from the elevator to the baths. Forty rooms with marble walls, tiled floors and porcelain tubs, for the regulation Hot Springs baths; separate rooms for Russian vapor, Turkish, needle, electric, and other cleansing and parboiling devices; hot rooms, cooling rooms; if there is anything in the "next to Godliness" business that is not to be found in perfection in this three-story palace of purification, it has not yet become known in civilized communities.

Next in order is the grand dining room—a model of elegance—extending the width of the entire building and the length of the main wing, with large windows occupying three sides and giving it that bright and cheerful appearance we have already noticed in the grand rotunda, and which, we will find before we get through.

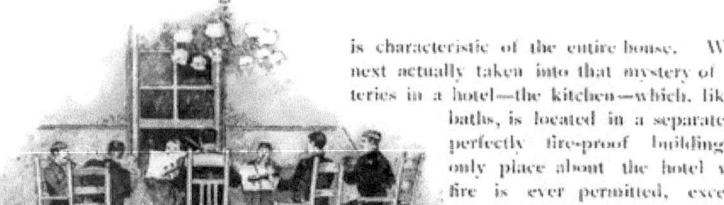

is characteristic of the entire house. We are next actually taken into that mystery of mysteries in a hotel—the kitchen—which, like the baths, is located in a separate and perfectly fire-proof building—the only place about the hotel where fire is ever permitted, except a glowing blacklog in the great fire-place in the grand rotunda on occasional chilly evenings. The secrets of the great range, the warming tables, etc., are disclosed to us, but we are heedless in our admiration of the absolute cleanliness on every hand. We cannot but think that any meal from this kitchen must be appetizing, and later we will try one, and find our pre-judgment correct.

Now we stroll back into the main building again, glance at the ladies' ordinary, the writing and card rooms, and then come to the grand parlor, a noble apartment, with royal furnishings, decorations and more windows. Then there is a smaller parlor for the ladies, which partakes of the same magnificence as the larger. Then come the guest rooms. We are shown one suite on each floor, and gloat over the handsomely carved oak furniture, the rich carpets and hangings, the cosy little closets and bath rooms, and the many other accessories, which have been provided for the delectation of the occupants. We finish the main building with a trip, via the elevator, to the grand observatory at the top of the building—an airy pavilion, 35 by 65 feet—walled with windows from which, apparently, the whole State of Arkansas can be seen, and containing chairs for a hundred people. What a grand lounging place for lazy people on a lazy day!

1—GRAND PARLOR. Park Hotel.
2—GRAND STAIRCASE,

The dancing pavilion is another building apart from the hotel, but connected with it by a covered way. This handsome structure is devoted exclusively to the votaries of Terpsichore and to other amusements. The Saturday night hop at the Park is a swell event at Hot Springs and the occasion of the gathering of the social clans of all the hotels.

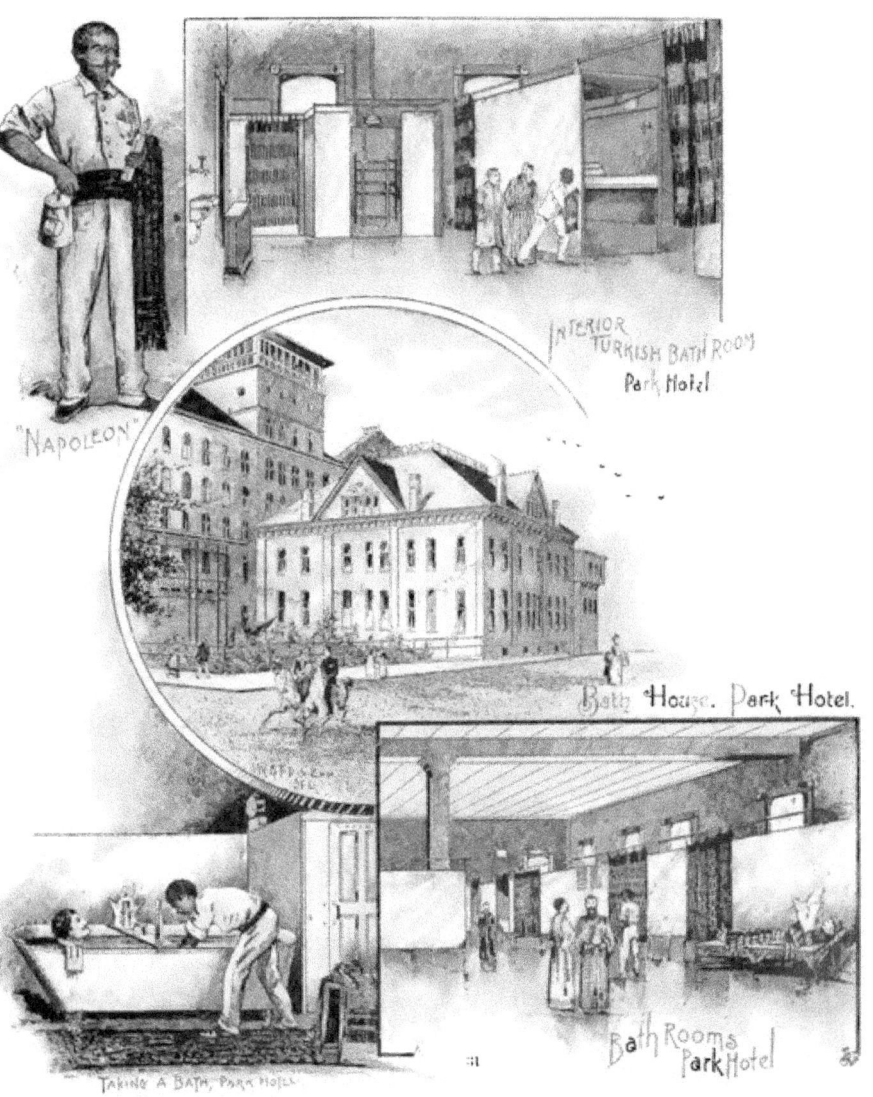

THE NEW ARLINGTON HOTEL.

IF you are an old timer at the Springs, you will be interested to learn that the Arlington as you knew it is a thing of the past, that the old structure has been demolished, and that a new, modern, magnificent hotel is to take its place, a hotel that will rival in all its features and appointments the luxuriance of the Eastman and the Park. All this is promised by the hotel company, and their uniform success in the management of the old property is the best possible assurance of the fulfillment of their pledges as to the New Arlington.

The site of the Arlington is unquestionably one of the very best at Hot Springs. On Central avenue, at the north end of bath house row, it is in the very heart of the city, and nearer to the Hot Springs themselves than any other hotel. The Hot Springs Mountain, from which all the springs flow, rises immediately behind the house to a height of three hundred feet, and the water from what has been considered the most efficacious spring for drinking and bathing purposes flows through pipes directly into the hotel building.

The new hotel will have a total frontage on Central avenue and Fountain street of 650 feet. The style of the architecture is of the Spanish *renaissance*, which is admirably adapted to the location and in striking contrast with that of other hotel structures in this city. It will be four stories in height, and built entirely of brick. The striking features of the front will be a fifteen foot veranda or arcade of brick arches, forming a continuous promenade of nearly 600 feet. The upper front will be relieved by balconies at suitable points. The two main corners of the building are emphasized by handsomely designed towers, twenty feet square, that extend thirty or forty feet above the roof, making excellent observatories, and adding to the general artistic effect. Special care has been given to the exterior to make it thoroughly artistic in proportion and design.

The interior of the hotel in arrangement, design and finish, will correspond in excellence with that of the exterior and will possess a number of features that will be marked improvements on others already built. The main floor will be raised six feet above the sidewalk line and will be sixteen feet in the clear. The others will be of proportionate height. Midway of the Central avenue front will be located the large rotunda, 53 x 86 feet in dimension. This will be the principal feature of the interior, and it will be a very striking one. In the rear, directly opposite the imposing entrance, the grand stairway will rise around the sides of a square to the floors above, under a circular glass dome, directly beneath the center of which will be placed, on the main floor, a fountain, altogether giving a very artistic and attractive appearance to this part of the interior. It will further be embellished by massive fire-places and mantels, paneled ceiling and fresco work. South of the rotunda is a large general parlor, 70 x 50 feet, which can be used for dancing. North of the rotunda is the gents' parlor, general card-room, waiting-room, etc. In addition to these, the hotel is provided with billiard-room, bowling alley, barber shop, bar-room, Western Union Telegraph office and all the latest conveniences. There will be 260 guest-rooms. These will be plastered with acme plastering, and the walls beautified by fresco penciling. The arrangement of the building has been such as to insure large, light and airy rooms. There are no dark rooms in the house, every

The Old Arlington.

one having outside light and ventilation. A commodious closet will be attached to each one, and the principal ones will have private bath, closet and fixed wash bowls. The furnishings will be in keeping with other appointments.

A fine passenger elevator will carry guests from first to fourth floors, and a large baggage elevator will insure quick transportation of baggage.

Ample fire escapes have been provided, by placing at the ends of each hall, on every floor, a large balcony constructed of iron, from which a walk will extend to the mountain slope in the rear, and from this the descent to the street can be made by walks without a single step. A complete return call system of bells and fire alarm will be one of the hotel's appurtenances.

The surroundings of the hotel will be made as handsome and attractive as skill can make them. Broad granitoid sidewalks and lawns will beautify the street front. The park and fountain on the north side of the building will be preserved and further adorned by a broad driveway extending around same from the *porte-cochère*, which will be at the Central avenue and Fountain street corner. The slope of the Hot Springs mountain, back of the hotel, will be laid out in walks and beautified

The New Arlington

with arbors and flower beds, providing a delightful retreat, and making pleasing the view from the rooms on that side of the hotel.

The elegant bathing establishment, which is to be erected on the site of the present New Rector, will be made a part of the hotel, being built by the same company, and can be reached from two floors of the hotel, through large, well-lighted and ventilated hallways. In external appearance, this will be a beauty in architectural design, and the interior arrangement is upon an entirely new, convenient and original plan, embodying features that are pronounced improvements, and making it the finest bathing establishment, in every particular, in this city, and second to none in the United States. There will be forty bath-rooms, and in addition there will be needle, shower and electric baths and all other appurtenances necessary to make it complete. The bathing department proper is an immense circle from which the bath rooms radiate on both floors. The series of cooling rooms or parlors, at different temperatures, are located in convenient proximity. Perfect light and ventilation are provided by a central light shaft, extending from the first floor to the roof, and terminating in a large glass dome with pivoted side lights. This insures direct light and ventilation to every bath room in the house. The walls and partitions of the bath rooms will be of marble, and the floors laid with Mosaic tile.

A steam heating and electric light plant of its own, will furnish heat and light for the entire hotel and bath house.

The observatory tower of the New Arlington cannot fail to be one of its most popular features. From the summit one will find spread out before him the most charming panorama imaginable. Directly opposite is the towering West Mountain, whose lofty heights are soon to be converted into a delightful park, with winding drives and foot-paths, fountains and flowers, and which even now is a wild, though somewhat rugged, wilderness, plentiful with attractions for the sturdy climber and explorer. Looking to the south, Central avenue stretches away until it loses itself in the pines of the Ouachita Valley, miles distant, lined on the one hand with the bath houses and on the other with the busy retail stores, whilst the broad roadway is gay with carriages and pedestrians. The towers of the Eastman, and the big cupola of the Park add picturesqueness to the scene, and the many other structures of varying size and architecture, combine to form a picture, the like of which is not to be found elsewhere on earth. To the north, Central avenue divides into Whittington and Park, almost at your feet, the former winding around West Mountain towards the setting sun, while the latter disappears among the hills to the eastward. In this direction the horizon is soon hemmed in by the mountains, but the view is nevertheless charming. At your back is the Hot Springs Mountain and the Government Park, while from the very *porte-cochère* of the hotel you can trace the beautiful Gorge of Happy Hollow, with its throngs of pedestrians, its crowds of merry laughing children, its donkey parties, its photograph galleries, and all its other "happy" accompaniments.

The New Arlington will continue under the management of S. H. Stitt & Co., who have done so much in the past to make the old hotel attractive and popular with patrons, and this management will be hailed as a guarantee of the prosperity and success of the new enterprise. Mr. Lyman T. Hay will occupy the position of manager, as during the past season, when he proved so popular a host, and will gather about him the best and most efficient corps of assistants to be found in the country.

Other Hotel Accommodations.

* * * *

AS before indicated, the visitor to Hot Springs is never at a loss to find a lodging place suited alike to his tastes and his pocket-book. Besides the three large hotels already described, and which are more especially patronized by the wealthier class of pleasure seekers and invalids, there are a dozen or more first-class hostelries at which lower rates prevail, and four or five hundred boarding houses. Of the hotels, among the best known are the Hotel Hay, the Pullman, the Avenue, the Waverly, the Hotel Worrell, the Josephine, the Sumpter, the Plateau and the Grand.

The more pretentious boarding houses assume names—the Albion, the Burlington, Taylor's, Magnolia Villa, Haynes Villa, for example. These houses are handsomely furnished, conveniently located, with pleasant surroundings, and are well patronized, as their excellent conduct deserves. Following the descending scale of prices, come the numerous lodging houses, whose only name is the invariable "elegantly furnished rooms for rent, with board," and last come the furnished rooms, nearly every house in the city, even to the humblest cabin, having "a vacant room" somewhere about the premises.

There is no good reason why the food served at any of the boarding houses should not be nourishing and of agreeable variety, and it is a commendable fact that in this respect Hot Springs enjoys an enviable reputation. A branch establishment of the Armour Dressed Beef Company furnishes good fresh meats in abundance and at as low prices as prevail anywhere. The recent development of Arkansas as a successful small fruit region, places these palatable and necessary adjuncts to a good table within reach of all, while there are numerous vegetable farms about the city, which contribute their quota to the gastronomic entertainment of visitors. The milk and butter are unusually good, there being several dairies, equipped with fine Jersey cows.

The fact that so many hotels and boarding houses of such diversified character exist and apparently flourish at Hot Springs, is in itself evidence, both of the enormous number of people who come hither in search of health, and also that no condition of wealth, or lack of the same, is a bar to the enjoyment of the beneficial effects resultant upon a use of the waters and baths. The water of the springs is thoroughly democratic.

Several of the most popular hotels and boarding houses are described in detail in the ensuing pages. Lack of space forbids special mention of all, but those named herein may be taken as fair examples of the rest. It is, perhaps, well to suggest that convenience to one's physician and to the bath house selected by the visitor is deserving of consideration in making a choice, as well as price to be paid and character of the accommodations.

The Hotel Hay, a cut of which is given herewith, has seventy-five guest rooms, with a capacity for 120 guests. Many of the rooms are *en suite*, making them desirable for families, and all are elegantly furnished and appointed with Brussels carpets, marble-topped mahogany and walnut chamber sets and wardrobes, the beds having curled hair mattresses on wire springs. The halls are spacious and well lighted, with high ceilings —each story being twelve feet high. The hotel is centrally located, within two blocks of Bath House Row, is heated by steam, has a passenger elevator and is lighted both by gas and electricity. A fine spring of chalybeate and magnesia water flows directly into the hotel and has been found very efficacious in troubles of the stomach, bowels, liver, kidney and urinary organs. The rates are from $2.50 per day, or $11 per week, upwards, and the service is unexceptional in every respect.

The Pullman is a fine brick structure located on Central avenue directly opposite the center of Bath House Row, and, for convenience of location, is not excelled at Hot Springs. The rooms are large, airy and well lighted and handsomely furnished throughout, and the table is furnished with the best the market affords. A café for ladies and gentlemen is one of the attractive features of this popular house.

The Hotel Worrell is also a finely appointed family hotel of the highest rank, located but one block from the Eastman and convenient to the depot, postoffice and principal stores. Its patrons are of the best social class, and they all speak in the most complimentary terms of the excellence of this popular house.

The Avenue Hotel is a large, handsome structure, located on Park avenue near its junction with Central avenue. The house has recently been enlarged and improved throughout, and it is one of the best furnished at Hot Springs. every room having marble topped furniture, dressing cases, wardrobes, etc. A new Hale hydraulic elevator, gas and electric lights. electric bells in each room, a Western Union telegraph office, are among the many conveniences with which the hotel is equipped. The bath rooms are exclusively for the use of guests and are in the hotel building, with a private hallway from the elevator landing to the ladies' bath rooms.

Bridges connect each floor with the mountain in the rear, affording easy egress in case of fire, and the stand pipes and fire hose, which can be operated instantly, make the hotel absolutely safe.

The Plateau, as its name implies, occupies an elevated position on Central avenue and commands a fine view of the valley, surrounding country and mountains. It is comparatively new, and neither pains nor money have been spared to make it thoroughly comfortable and convenient. The ventilation and drainage are perfect; it is furnished with elevator, electric bells and lights, automatic fire alarms in each room, and the service is first-class. The location is central, and especially so for commercial men, as the heaviest business houses are within two blocks of the house.

The rates are $12 to $20 per week, according to location of rooms, the hotel being open the entire year, and rates being lower during summer and fall than during the winter and spring.

Avenue Hotel.

The Plateau.

The New Waverly

is one of the very best of the smaller hotels, and is noted for the high standard of excellence maintained in all departments by its management. It is an attractive appearing house, with numerous verandas, good sized and richly furnished rooms, and is located on Park avenue but a short distance above the head of Central avenue. All the modern conveniences—passenger elevator, electric lights, electric bells, etc., are provided, and the service is unusually good. The Waverly is essentially a family hotel, and caters only to the very best class of people. The table is one of the best in the city, and well supplied at all times with seasonable delicacies. Closed carriages are placed at the disposal of guests free of charge, for conveyance to and from the bath houses, which are located about two ordinary city blocks from the hotel.

The Sumpter House

has been so thoroughly changed by additions, alterations and rebuilding, that its old friends and patrons would scarcely recognize it. It has been newly furnished throughout, and is now one of the best houses at Hot Springs. Some of the rooms in the new extension are equal to anything to be found in the city. Its location on Exchange street, within 200 feet of Central avenue, makes it most desirable for visitors who prefer to be where they are convenient to the baths and the main business part of the city.

The Hotel Josephine is situated at Whittington avenue and Cedar street—one of the best locations in the city—is a new house, with sixty rooms newly furnished throughout, and accommodates about one hundred people, its guests being largely composed of families. The house is kept so nicely and everything about the premises is so neat and tidy, that it is frequently known as "The Band-Box House."

The Albion, Magnolia Villa, Taylor's, Haynes Villa, and many others, may all be classed as family hotels of the better class. After them come the boarding houses, of all sorts and conditions, and as numerous as leaves in Vallambrosa. Hot Springs is the *ne plus ultra* of the boarding house keeper, and the native population seems to be largely made up of this ubiquitous class. Nearly every house in the town bears some legend, often grotesque as to both orthography and the formation of the letters, announcing "One furnished room to let," "Boarders wanted," etc., etc. One sign, at the foot of a rugged mountain which even the nimble goat would find troublesome climbing, announces, "One furnished room to let with cooking up hill," a variety of the culinary art that is doubtless indigenous to Hot Springs.

Speaking of signs, those seekers for the curious who delight in hunting up inscriptions, ill-writ, mis-spelt, ludicrous and startling, will find a prolific source of amusement in the signs at the Springs. One rather comprehensive merchant in the suburbs advertises "Sausiges sold hear; also Cigars, Ice Cream and Drug Store." Another affects the alliterative, after the following sibilant style: "Sweet Sider, five sents a skooner." Another has "Oranges, pea-nuts, summer drinks and Book store." A necromancer on Malvern avenue announces "Fortunes told bear, past and futher" —an unique future as far as the sign is concerned, surely.

The rebus is a favorite form of sign advertising. The ark, can and saw are to be seen everywhere, even on the handles of souvenir spoons. One of the leading jewelry stores is known as "The Can," and is decorated with an immense teapot. It might be as well to say here that the can is the utensil most in use at Hot Springs, and on a pleasant morning one may see a thousand people on Central avenue, bearing tin teapots either filled or about to be filled with hot water.

I must not forget one pictorial sign nailed to a tree beside one of the carriage roads. It has on one end a picture of a winged man, clad in a crown and a night gown, who is soaring through a Prussian blue sky, with a satisfied smile on his face and this inscription beneath his feet: "This man took Carter's Compound Cure." On the other end is a harrowing depiction of a man writhing in a huge pot of flames, while his Satanic Majesty prods him vigorously with a pitchfork. Here the legend informs you that the sufferer did *not* take Carter's Compound Cure. Mr. Carter has evidently discovered an excellent *post-mortem* remedy.

The Springs and Their Properties.

THE hot springs, the loadstone which has drawn together these thousands of people from all parts of the world, are seventy-one in number, with a temperature ranging from 76° to 157° Fahrenheit, and a flow of half a million gallons daily. The cause of their marvelous medicinal effect is still a mooted question among physicians and chemists. Careful analysis by eminent specialists show that, on an average, the waters contain 12.94 grains of material in solution to the gallon. Of this material, nearly sixty per cent is carbonate of lime, over twenty-one per cent is silica, nine per cent is carbonate of magnesia, while the remainder is chiefly chloride of sodium (common salt), sulphate of soda (Glauber salt), and sulphate of potash. This is but a slight proportion of minerals, in fact, no more than is to be found in many springs and well waters used for domestic purposes. It is, therefore, an accepted theory with most practitioners that the wonderful virtue of the waters lies in their natural heat, which seems to possess peculiar, perhaps magnetic qualities, not characteristic of those of other warm springs, or of waters artificially heated.

Dr. William Elderhorst has this to say regarding the curative qualities of the water: "In many forms of chronic diseases, especially, its effects are truly astonishing. The copious diaphoresis (perspiration) which the hot bath establishes, opens in itself a main channel for the expulsion of principles injurious to health.

Glimpse of Bath House Row, on Government Reservation.

made manifest by its peculiar odor. A similar effect, in a diminished degree, is effected by drinking hot water—a common, indeed, almost universal practice among invalids at the Hot Springs.

"The impression produced by the hot douche, also, is indeed powerful, arousing into action sluggish and torpid secretions; the languid circulation is thus purified of morbific matters, and thereby renewed vigor and healthful action are given both to the absorbents, lymphatics, and to the excretory apparatus—a combined effect which no medicine is capable of accomplishing.

"The large quantity of free carbonic acid which the water contains, and which rises in volumes through the water at the fountain of many of the springs, has undoubtedly an exhilarating effect on the system, and it is, no doubt from the water of the Hot Springs coming to the surface charged with this gas, that invalids are enabled to drink it freely at a temperature at which ordinary water, from which all the gas has been expelled by ebullition, would act as an emetic."

The moon was now, from Heaven's steep,
Bending to dip her silver urn
Of light into the silent deep;
And the young nymphs on their return
From those romantic glens, found
Their other playmates, ranged around
The sacred spring, prepared to tune
Their parting hymn, ere sunk the moon,
To that fair fountain, by whose stream,
Their hearts had form'd so many
a dream.

All of the Hot Springs, with one exception, flow from the Hot Springs Mountain, on the east side of Central avenue. Formerly, the hot water, with its accompanying clouds of vapor, could be seen issuing from the ground; but it is now, for the sake of economy and cleanliness, piped from the various springs to a common reservoir, whence it is distributed to the different bath houses. This collection and

The City of Hot Springs.

DAME Nature planned the City of Hot Springs after her own unique methods. She planted fountains of living waters in the hills, and made them her engineers. With the mighty power of persistent gentleness she chiseled out the avenues from the stubborn flint and jasper, and shaped the contour of the coming town. With her invariable abhorrence of straight lines and right angles, she wound her streets in sinuous sweeps around the bases of the mountains, made tortuous paths to their summits for the horseback rider and pedestrian, shaded hill and valley alike with her noblest and choicest forest trees, provided suitable sites for hotels, business houses, residences and bath houses, arranged with the lower regions for the necessary privileges, and turned

BIRDS EYE VIEW OF HOT SPRINGS 1891

1. View near Race Track.
2. Live Tree Bridge.
3. Near Cemetery.
4. The Cascades.

on the hot water. Man and the United States government have done the rest. They have been pegging away at it for sixty years, and the result is the present thriving, picturesque city of 15,000 souls, with its magnificent hotels, busy banks and stores, costly churches and fine residences.

It will be seen from the accompanying cut that there is little regularity about the streets and avenues. The main thoroughfares follow the courses of the streams between the mountains, Central avenue, the principal business street being a narrow valley running north and south between the Hot Springs and West mountains. This was formerly the bed of Hot Springs Creek and was filled with huge boulders which, with the wanton course of the stream, made the valley well-nigh impassable. The government work of confining the creek to a tunnel or underground passage, clearing the valley and constructing a broad well-paved street above, was an engineering feat of no small proportions. It was made necessary, however, before the bath houses could be constructed or a business street established.

On Central avenue are located all of the bath houses, except the Park, and nearly all of the hotels and business houses of the city. The hotels are mostly at the upper end of the avenue, except those built in the past few years, and the business houses are down toward the lower end and around its entrance to the larger valley. Of late years, the avenue having been more closely built up, some of the business portion of the town has gone out into the Ouachita Valley, which now claims some fine streets and business blocks. The bath houses occupy about three blocks on the

No. 1. GRAND OPERA HOUSE.   No. 2. POST OFFICE.

east side of Central avenue, near the heart of the city, in front of Hot Springs Mountain, from which they are supplied with the thermal waters.

The Post Office building is a finely to the hotels and stores. The Opera House is also on Central avenue, near the north end, and is a handsome brick building; completely equipped in a theatrical sense, and occupied during the season by the best opera and dramatic companies.

The business part of the city

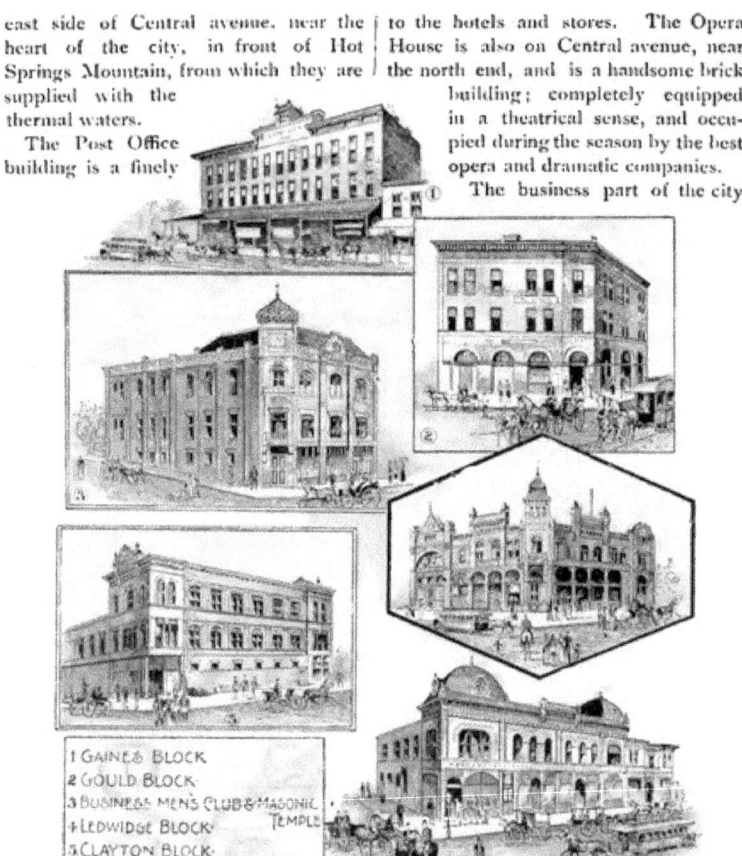

1 GAINES BLOCK
2 GOULD BLOCK
3 BUSINESS MEN'S CLUB & MASONIC TEMPLE
4 LEDWIDGE BLOCK
5 CLAYTON BLOCK
6 METROPOLITAN BLOCK

appearing brick structure of three stories, located on Central avenue, convenient is of a substantial and permanent character and shows decided improvement in the past two or three years. Several solid business blocks have recently been, and are now being erected, and still greater changes may be looked for in the near future. Stores and shops are well stocked with everything calculated to contribute to the material well-being and happiness of visitors and residents. Everything sold elsewhere can be obtained, and some things that are not—namely, Hot Springs diamonds made from quartz crystals, agates, and other beautiful stones found in the vicinity and fashioned into many curious and useful shapes, and articles by local lapidaries.

No city in the land can boast of a retail row, that, for attractiveness, variety and oddity, can compare with the business side of Central avenue. As a charming young Miss at the Eastman put it, "It's better than going to the circus." Jewelry stores, book stores, dry goods stores, saloons, drug stores, pool rooms, restaurants, furnishing goods stores, grocery stores, more jewelry stores, doctors' offices, more drug stores, shooting galleries, more saloons, and so on and so forth—each with a distinct picturesqueness of its own—until one almost wearies of the endless, and, in many cases, striking contrasts. Still, this is Hot Springs, and no one would have it otherwise. It would rob the city of half its charm to change Central avenue into a row of stiff, formal brick blocks with staring plate-glass windows.

There are three substantial banks in the city, three daily papers, and the little folks are provided with an excellent school system.

Street railway facilities are remarkably good, there being nearly eight miles now in operation, connecting all avenues and sections of the city via Central avenue. The rolling stock is new and the cars will compare favorably with those in use in larger cities. A street car line from the Arlington up Happy Hollow is the latest improvement in this direction. While mules are at present the motive power, it is contemplated that a change to electricity will be made in the near future.

Hot Springs enjoys an excellent supply of pure water for drinking purposes and domestic uses, and it is furnished in such volume as to be of effective use in case of fire.

A clear mountain stream, fed by huge springs, about two miles north of the city, was made to form a lake half a mile or more long by means of a dam of solid masonry, thirty-eight feet high, extending from mountain to mountain. From the lake this water is forced into an immense reservoir on the summit of the mountain, 250 feet above the streets of the city. This head gives so great a pressure that a stream from the largest hose can be thrown over the highest buildings without the aid of fire engines. This improvement cost the city over $150,000, and is capable of furnishing 2,250,000 gallons daily. The fire department is well organized and equipped with all necessary paraphernalia, and shows, when occasion requires, that it is fully prepared to fight the fire fiend successfully.

Hot Springs Water Works

Neptune Fire Company

Water Works Dam and Reservoir

1 CATHOLIC CHURCH
2 M·E·CHURCH SOUTH
3 FIRST BAPTIST CHURCH
4 PRESBYTERIAN CHURCH
5 St LUKE'S CHURCH, EPISCOPAL
6 COL·M·E·CHURCH
7 URSULINE CONVENT
8 SYNAGOGUE

Hot Springs Churches

The city is well supplied with churches, nearly all the leading denominations being represented, and all in a flourishing condition. Some of the church buildings are handsome specimens of architecture, St. Luke's Episcopal, on Rector street, directly opposite the Eastman, being the most pretentious. The large stained glass window in this church is one of the handsomest in the West, and the entire structure is a model of neatness and beauty.

The choice residence portions of the city are on Park and Whittington avenues, though there are many handsome homes in the southern section, especially on Malvern and South Central avenues, and the Government has erected expensive and beautiful houses on the reservation for the superintendent and surgeon in charge of the Government hospital. As Park and Whittington avenues lie at the bottom of valleys with lofty mountains on either side, the residences are necessarily built on sloping ground. The mountain sides are thickly wooded, and the effect of a cozy cottage or mansion nestled among the native forest trees, far above the street level, with grassy terraces leading down to the entrance gate, is both unique and picturesque. Many of the residences of Hot Springs will compare favorably, in elegance and beautiful surroundings, with those of larger cities, and are pleasing evidences of the wealth and refinement which have found their way to this city of the Ozarks. The accompanying series of cuts give one a fair idea of the homes of Hot Springs, a feature of which every resident is justly proud.

A number of northern people, appreciating the advantages of living under one's own vine and fig tree while at Hot Springs, have purchased lots and erected cottages, which they occupy during the season. This plan has met with great favor, and the pioneers have many imitators from all

parts of the country. The sanitary conditions for a residence at the Springs are absolutely perfect. The drainage system could not be improved upon; the atmosphere is healthful and bracing, and the water pure and wholesome. As stated elsewhere, the markets are as well supplied, and the prices of meats, vegetables and groceries as reasonable as can be found anywhere. There are good schools for the children, good churches, plenty of amusements, excellent mail, express and telegraphic facilities, and emphatically no lack of medical attention or drug stores. The retail stores are of all kinds, and any article of personal use, comfort or luxury can be purchased.

In regard to the healthfulness of Hot Springs, a most important consideration to one contemplating a residence there, government statistics show that out of a total of 486 cities and towns in the United States, only five have as low death rates as Hot Springs. There is only one city in British America having one as low, and none at all in England or Central Europe. The death rate per one thousand inhabitants in 1889 was 7.74 at Hot Springs. When it is taken into consideration that a large percentage of the citizens of Hot Springs went there afflicted with disease,

1. Major G. G. Latta's Residence.
2. Dr. E. C. Ellis' Residence.
3. Dr. Pollard's Residence.
4. Dr. Passmore's Residence.
5. Entrance to Dr. Passmore's Res

and, being cured, have since made it their home, this percentage seems all the more remarkable. In connection with the great healing powers of the springs it is almost miraculous. The two go hand in hand, and it would seem as if Nature had left nothing undone to make this the great sanitarium and retreat to

1. Major Gaine's Res. 3. Mr. C. W. Baxter's Res.
2. Dr. Greenaway's Res. 4. Mr. C. N. R 's Res.

ward off disease and prolong life. The death rate previously quoted applies only to the permanent residents of Hot Springs, numbering some eighteen or twenty thousand. The rate among the fifty thousand annual visitors is very small,

1. Judge Kimball's Residence.
2. Dr Garnett's Residence.
3. Dr. William's Residence.
4. Dr. Ellsworth's Residence.

being only a little more than one and one-tenth per cent, and in nearly all instances of death among visitors, the disease causing the same had reached such an advanced stage that recovery, even under the most favorable conditions, and with the best medical attention, was an utter impossibility.

Happy Hollow.

WEST MOUNTAIN FROM HAPPY HOLLOW

Happy Hollow, from West Mt.
Government Reservation Office

EVERYONE who visits Hot Springs, quickly succumbs to the charms of Happy Hollow, and becomes its faithful admirer. A pleasant valley it is, indeed, and the favorite resort of pedestrians, though a street car line has recently invaded the picturesque glen; much to the disgust of many who consider the new-comer an unwarranted encroachment upon their pet promenade. Happy Hollow has, in a general way, been so happily and graphically described by Mr. Wm. L. Belding that his description is herewith inserted. He says:

"This is a dell that the imagination of the Greeks would have populated with all sorts of superhuman beings. It would have been a kingdom for fairies, a favorite haunt for nymphs and *dryads*, and might have been a trysting place for the gods and goddesses themselves.

"The place is neither a gorge nor valley, but a quiet and peace-inspiring glen. A narrow roadway hewn from the side of the mountain, which disputes possession of the bottom of the dell with a creek. And a most delightful road it is, with the mountains thrusting their feet down from either side and nearly crushing you beneath the rocks, and rising gracefully to the height of three or four hundred feet, their sides covered with huge rocks and tall sighing pines and

Scenes in Happy Hollow.

oaks, which, in the autumn, form a most beautiful picture in yellow and emerald. The foot path is excellent, the dell is cool, and there is a new charm at every step.

"At the terminus of the carriage road is located the celebrated Happy Hollow Spring. It is not hot water that boils up in the unique little summer house that has been built over the spring, but, notwithstanding that, it is a mineral water of high grade and possessed of great medicinal virtue. It is used solely for drinking purposes and in connection with the hot baths. Dyspepsia and indigestion have no show whatever when brought into contact with Happy Hollow water. It drowns rheumatism and gout, and, as it acts directly on the liver and kidneys, it purifies the blood, producing a beautiful complexion, and is almost a specific in all diseases of the urinary organs. It is visited by great numbers daily who come to drink the water. Beyond the spring the road ceases, the glen becomes steep and the path is blocked and turned from side to side by huge boulders and the jagged, projecting edges of the mountain. At times it is almost impassable, and the explorer would like to give up and turn back, were it not that he desires to pursue the rocky way to the end and see where it commences or terminates."

Mr. Belding's explorer should not be discouraged. If he persists in pursuing the "rocky way," he will be amply rewarded for his climb, as the path leads him to the summit of Hot Springs mountain, with its picturesque walks, grand views, beetling cliffs and magnificent forest. The walk along the crest of the mountain to and down the Grand Boulevard to the Government hospital, is one of the greatest attractions at the Springs, and should not be overlooked by anyone.

Happy Hollow is the lair of the Hot Springs burro, a large drove of these interesting little brutes being kept on hand for the use of pleasure parties. On their backs the intricacies of the upper glens may be readily explored, and, while one does not present a very imposing appearance mounted upon the diminutive animals, they are a source of constant enjoyment, not to say of uproarious hilarity.

The photographer of Happy Hollow reaps a rich harvest from these donkey parties, who send thousands of pictures of their mounts to the folks at home.

The Man on Horseback.

HOT SPRINGS is a paradise for the equestrian. To every point of the compass, beautiful shaded roads and bridle paths meander away over mountain and plain, through wooded dells, and across sparkling streams. No American resort can boast of so many charming drives or of such infinite variety. One can take a different route every day for a month and not exhaust the repertory, and find each day some new and unexpected charm. A map will be found inserted at the back of this book, showing the location of the various roads and trails, and by its use and guidance the entire rugged and picturesque vicinity may be explored and enjoyed. The numerous mineral springs and creeks are also located on this map.

The horses to be hired at the Springs are of unusual excellence, and a source of agreeable surprise to the visitor. They are mostly Kentucky stock, highly bred and especially trained for horseback riding, gentle, sure-footed, speedy and of easy gait.

One of the most delightful drives is out Park avenue. The beginner will find the ride around North Mountain, through the gorge and returning by the railroad depot, one of sufficient length and beauty, to start with, and also to warrant an investment in Pond's Extract or some other alleviator of soreness upon his return. Or he may go to the big Chalybeate Spring, some two miles out, one of the features of the locality. Here, a spring of clear, sparkling water, strongly impregnated with iron, and some eight feet in diameter, gushes out of the bank in a stream the size of a brook, and tumbles in a miniature cascade into the branch of the Gulpha that murmurs alongside. The water has many virtues, and is considered extremely beneficial by the thousands who use it. The spring is in a charming little valley, surrounded by giant oaks and other trees, and is a delightful retreat on a warm afternoon. Across the road on the hill-top is a long rambling one-story structure, formerly used as a hotel, which lends novelty to the scene; and as the visitor looks, a little black-eyed, bare-footed nymph, glass in hand, darts from the

door, and, dancing across the valley to the spring, stands ready to serve the water to the thirsty equestrian. A little farther on the road forks, the one on the right leading through a fertile valley, across the south fork of Saline river, on to Little Rock, while the one on the left winds around the mountains and through the forests to Mountain Valley, a popular health resort, of which more will be said later on.

Another charming drive for an afternoon, is around Sugar Loaf and West Mountains, going out Whittington avenue, climbing and descending the mountain by a tortuous road, which at every turn displays some new and magnificent view of valley and distant mountain range, thence to and across Bull Bayou, down the west bank of this beautiful stream, to

the Bear Mountain road, and thence back to the city. The ride to the Ouachita by any one of half a dozen roads is always a pleasant one, as is that to Gillen's White Sulphur Spring and to Potash-Sulphur Springs, described further along.

Horseback parties are extremely popular, and any pleasant afternoon squads of both sexes may be seen dashing off to the mountains, to return at dusk, with faces all aglow with health and with magnificent appetites for dinner.

Carriage Drives.

While the country roads, with a few exceptions, are a little rough for carriage driving, the drives around the city are good and in sufficient number and variety. Most of the streets and avenues are macadamized, and improvement in this direction is going on constantly. It is understood that a good portion of the $74,000 realized at the recent sale of government lots will be devoted to the construction of boulevards and carriage roads. As it is, the boulevard on Whittington avenue, the Grand Boulevard, the road to Potash-Sulphur Springs, Park, Malvern and Central avenues, and the Dallas and Arkadelphia roads, afford all needed facilities.

Any kind of a vehicle can be hired, from the most elegant landau down to the buck-board.

Potash-Sulphur Springs.

A SHORT mile from Lawrence station, and seven miles from Hot Springs, are located the health-giving Potash-Sulphur Springs. Ten trains pass Lawrence daily, and are met by hacks, which convey visitors to and from the springs. A favorite way of reaching them from Hot Springs is on horseback, the ride being one of the most picturesque in the country. A handsome two-story hotel, elegantly furnished, and numerous cottages afford the best of accommodations to guests, and the table compares favorably in excellence with the leading hotels of Hot Springs. The air at Potash-Sulphur is at all times pure and balmy, while surrounding mountains and valleys afford picturesque scenery, and charming walks and drives. Deer, wild turkey, quail and other game afford good shooting, and the Ouachita river, one mile distant, furnishes abundant sport for the angler.

The springs, which are near the hotel, are five in number, but all possessing similar properties. The waters are sulphureted alkaline, and are highly esteemed by the medical fraternity of Hot Springs, who send many of their patients there to spend a few days while resting, after taking a course of hot baths — and always with beneficial results.

Dr. John C. Branner, State Geologist of Arkansas, who has made a careful examination of the mineral characteristics of these waters, says of them:

"The importance of the Potash-Sulphur waters is too well known to admit of question, and so long as the waters do what is claimed for them, it makes but little difference whether their virtues come from one mineral constituent or another. The chief ingredients of the water are sodium sulphate, sodium carbonate and potassium chloride. The sodium amounting to 13.66 grains per gallon, the potassium to 3.51 grains per gallon. The three springs vary in strength, that on the east side of the branch, and nearest the bowling alley, carrying the largest amount of solid matter in solution. This latter has 66.00 grains to the gallon; the spring south of it, on the same side, contains 49.68 grains to the gallon, while that on the west side of the stream contains 31.02 grains to the gallon."

The following diseases are cured or benefited by the use of the waters of this valuable spring: Dyspepsia, gout, rheumatism, affections of the liver, kidneys and urinary organs, female diseases, dropsy, and all complaints originating from an excess of acid in the system, skin diseases and chronic dysentery. In diseases of the kidneys and urinary passages, stricture, gleet, and especially in calculous affections, there is no known remedy so efficacious as this water. It acts as a solvent in the various forms of gravel, and is very efficacious in the treatment of all mercurial diseases. For many of these diseases the Potash-Sulphur water is considered the best.

Gillen's White Sulphur Spring.

GILLEN'S is another favorite resort for the visitor to Hot Springs, whether he be sick or well, and he goes there in crowds, in carriages, horseback, and even on foot. The distance is in the neighborhood of three miles, and a rugged, rambling, romantic three miles it is too. Leaving the city, the way leads at once into the forest along the south side of Hot Springs Mountain. The fording of the Gulpha is the first startling diversion—particularly, if that obstreperous little stream is "up," as the local vernacular puts it. An old mill in a state of picturesque dilapidation stands near the ford, and a rude cabin nestles under the pines close by, with the inevitable Arkansas accessories of dogs, pigs, chickens and youngsters. Beyond the ford the road takes up a tortuous winding and twisting around the hills, through swamps and thickets of scrub oaks, occasionally crossing a clearing, then skirting a noisy mountain stream, giving frequent glimpses of lofty mountains and deep gorges, until, after a final sharp curve, it brings up at the entrance of the hotel.

Like the other resorts in the vicinity of Hot Springs, Gillen's has every charm that forest and stream, mountain and valley, can lend it. The hotel is a two-story frame building, well enough furnished and managed, and much resorted to by parties of Hot Springers, who ride or drive out in the afternoon, take supper, and return in the evening.

The spring occupies a fairy-like grotto of rock in a pagoda located in the center of an enclosed park across the road from the hotel. The waters are of the white sulphur variety, as indicated by the name, but are wholly free from sulphureted hydrogen, and therefore unusually palatable for sulphur water. Indeed, the taste smacks more of the chalybeate than of the sulphuric. The waters contain carbonate of iron, lime and magnesia, and very small quantities of sulphuric acid, and of free carbonic acid. There is no trace of chlorine. When exposed to the air, a small amount of iron oxide is slowly deposited. The total mineral solids per gallon are sixteen grains.

The white sulphur water is employed with beneficial results in all cases of dropsy, liver and stomach disorders, diseases of the kidneys and bladder, and, like the potash-sulphur, is tonic in its nature, and counteracts the weakness incident to a course of hot baths.

Gillen's Spring is a good starting point for several of the local natural wonders, notably Hell's Half Acre and the Thousand Dripping Springs. The former place of Sheolian name is distant but half or three-quarters of a mile from the hotel, and easily accessible, the landlord tells you, always, provided "the creek is not up." This uppishness, however, is a matter of such frequent occurrence that one often is inclined to wonder why they do not dam the creek—and proceeds to do so for them. The distance may be covered on foot or on a horse, the former method being preferable for many reasons, and is a pleasing though somewhat wearisome bit of mountain climbing. Once arrived, however, the visitor is well repaid for the undertaking.

Imagine a tract of an acre or more, sunken from ten to thirty feet below the level of the surrounding territory, and presenting nothing to the view but a jagged, jumbled, chaotic mass of sharp-edged, irregular, multi-colored rocks. Rocks of all sizes and shapes and compositions; rocks of limestone, slate, flint and granite; rocks igneous and aqueous; and rocks the like of which are not to be found elsewhere. A barren, weird, forbidding conglomeration of boulders, an arsenal for Titans. Not a blade of grass, not a shrub, not even a lichen dares brave the atmosphere of death and desolation which

seems to pervade the uncanny spot. Yet the lofty pines, the spreading oaks, the young undergrowth of shrubs and the wild flowers flourish even to the very brink. Indian tradition has it that, when Gitchee Manito, the great spirit, smote the crags of the mountains and released the imprisoned hot waters for the healing of the nations, he, finding no suitable place for the disposition of the shattered fragments, thrust his mighty finger into the earth and dumped them in the hole.

Another local legend, from which the name undoubtedly comes, is that the spot is nothing more nor less than the original bottomless pit, with, paradoxical as it may appear, the Old Nick himself at the bottom, where he lies chained and under the crushing weight of innumerable boulders, powerless to do aught but groan and curse his awful fate. The old settlers who whisper this blood-curdling story among themselves, corroborate it by tales of strange experiences of belated hunters caught near Hell's Half Acre at sundown. It is asserted that time and again have they inhaled the sulphurous breath of the imprisoned demon as it rose from amongst the rocks, and have heard deep underground moans of pain and shrieks of savage profanity.

As either legend solves the mystery, the reader may take his choice. The fact remains that the spot is a strange freak of nature and well worth a visit.

The Thousand Dripping Springs, another natural curiosity, are located about a mile and a half to the northeast of Gillen's and can be reached by a fairly good road. They issue from a huge rocky ledge which overhangs the roadway, and which is pierced by a myriad of crevices, each one forming a separate spring. The little cascades unite at the foot of the ledge, and form a stream of considerable size which dashes across the road into the woods and soon loses itself in the omnipresent Gulpha.

There is one advantage about the trip to Gillen's. One can return to Hot

Gillen's. The writer could never find it in the wide, wide world.

During the past season Gillen's has had an unprecedented custom, and at times the hotel accommodations were wholly inadequate. To obviate this difficulty in future, Mr. John Gillen, the proprietor, contemplates building several neat cottages for the use of families, and otherwise improving the property. He also intends to improve the road from Hot Springs, with the idea of making it the best carriage and horseback drive out of the city. These commendable projects accomplished, Mr. Gillen can congratulate himself upon possessing a magnificent health resort, and one that would be very attractive if entirely independent of its greater neighbor, Hot Springs.

Springs by a different route. The great trouble with the roads of this vicinity is that they all go off in different directions, and attend strictly to the business of reaching their destinations. There are no intersecting or cross roads except in a few cases, the one referred to being an example. To any observant equestrian it is rather monotonous to retrace one's steps over the same road used in starting out, and the ability to vary the returning route lends a pleasing variety to the day's exercise. The landlord will explain all about the other road from

The hotel at Gillen's is named "The Victoria," and it is open the year round. Rates are from $10 to $12 per week. The rooms are large, well furnished and ventilated, and the table excellent. A hack line runs daily between Hot Springs and Gillen's, making the journey in about forty minutes.

Mountain Valley Springs.

THE Mountain Valley Springs are located among the Ozarks, about twelve miles from Hot Springs, and the trip affords an excellent opportunity to become acquainted with Nature's grand handiwork. The journey can be made any way to suit the inclination of the tourist. There is a regular hack line running daily from the hotels of Hot Springs to Mountain Valley. There is the more private and independent way of hiring your own horse and buggy, and driving and stopping at will, wherever you are interested, or it can be made the objective point of another horseback tour. It is one of the most charming rides in the vicinity. The road is over the mountains, plunging into gorges and through enchanted glens. Notwithstanding it has been recently improved, it is still a mountain road, and at times leads by points of thrilling interest. This very enjoyable ride terminates in one of the most beautiful valleys in the Ozark range. On all sides are the mountain peaks, shutting in a portion of the earth of paradisiacal beauty, peace and quiet. Tall primeval forest trees are singing Nature's hymns over the valley. The patches of open are verdant with grass and shrubbery, and an air of seclusion and rest is all-prevalent. You will scarcely expect to find here a modern six-story hotel, with "all the latest conveniences and comforts," as the hotel men say—and you don't. The hotel is entirely in keeping with the surroundings. You can expect quiet and rest at an inn, but not at a modern resort hotel. This is just the difference between the big, bustling hostelries of the city we have just left behind, and the low, quaint structure nestling in Mountain Valley. Its picturesque architecture is Southern in character, being one story in height, and covering a great deal of ground. Cool, spacious verandas cover the whole front of the hotel. Trailing vines overrun it in greatest profusion, and, altogether, it is a charmingly rare picture, and fills the beholder with an overpowering desire to settle down here, bag and baggage, and stay until satisfied with Nature's charms.

Nature sometimes scores a bull's-eye in her arrangement of things, and this feat was accomplished when these springs of health-renewing mineral waters were caused to burst forth in the midst of this beautiful valley. It is a matter of regret that the whole valley and springs could not have been located at a greater distance from the Hot Springs, so that they could have a chance at fame on their own merits. The waters are very valuable, used in connection with the hot baths, as their combined minerals constitute a curative medium for internal use not equaled. The analysis shows a large percentage of bi-carbonate of iron, lime and magnesia, sulphate of lime, chloride of iron, chloride of iodine, phosphoric acid, but not even a trace of organic matter.

The Ouachita River and its Inhabitants.

THE goal of many a horseback ride, the joy of the artist, the delight of the fisherman—the Ouachita occupies no mean rank among the attractions of Hot Springs. It is a child of the mountains, and is nourished by hundreds of brooks and springs that flow from their maternal bosoms. It sweeps in an irregular, sinuous semi-circle through the western and southern portions of the county, ranging from five to ten miles distant from Hot Springs. It may be reached by any one of six or seven roads. All are equally attractive to the equestrian, but the fisherman might not be so easily satisfied.

The writer made promise of a fishing expedition early in this book, at the time we caught that first glimpse of the river from the car window, and now that we have "done" the city, the baths, the hotels, and all the neighboring resorts, it is time the promise should be fulfilled.

We will wait until there has been a rain storm before we tackle the bass, for these fish are exceedingly shy during dry, pleasant weather, and take the hook best the day after a down-pour. The rainy day arriving, we get our outfit in readiness. You want a pair of wading boots, if you are not fond of wet feet, for this Arkansas fishing does not consist of sitting on the bank, or on a log all day and dropping your line in one place. If you want fish, you must go and work for them. A ten or twelve ounce bass rod, a good oiled silk line, some bass hooks on gimp, and some split shot, are all the tackle necessary. You will find plenty of darkies in the city who will supply the minnows, drive you to the river, and, even—as they do in many cases—catch your fish for you. The minnows caught in the Hot Springs Creek, above the Springs, are the toughest and most tenacious of life I have ever seen anywhere, so, when we start out at sunrise, our equipment is perfect. We can go to Thornton's mill, and fish below the dam. We can go out the Wildcat road, and boat it down the river for miles. We can strike the stream at Brown's ferry or Hurtz's ferry, or we can go by rail to Cove Creek. At any point, with conditions favorable, we can find fish, and we catch them too. We put in the day, or the two or three days, and when we return we will find, with ordinary luck, that we have a goodly assortment of large-mouthed black bass, of jack salmon, of black perch, of sunfish, and an occasional blue cat. The Ouachita fish are gamey, of firm flesh and excellent flavor. The best bass fishing is from March 1st until the last of June, when they begin to spawn.

There is no fishing in the tributary streams if the Ouachita is low, but when the latter gets on the rampage, as it frequently does, large schools of bass, salmon and cat, run up the Gulpha, Bull Bayou, Cove and Hot Springs Creeks, for sometimes a mile or two, and then the fisherman can get all the sport he desires.

Fishing and hunting are so closely allied that it will not seem amiss to state at this juncture that the hunter can find no better field for his prowess than the mountains in the vicinage of Hot Springs. Deer are plentiful, wild turkeys, quail and smaller game are abundant, and if he can get on good terms with an owner of hounds, he can enjoy a genuine fox hunt, there being numbers of gray or silver foxes in the forest glades of the Ozarks.

Hot Springs in Legend and History.

The legends of a locality lend it a charm and give it drawing powers independent of everything else. The mysterious phenomenon of hot water issuing from the ground naturally filled the poetic Indian imagination with strange fancies, and was a source of much speculation to his aboriginal mind, resulting in many legendary tales. Tradition is not wanting, either, connecting the Hot Springs with the Fountain of Youth persistently searched for by Ponce de Leon and De Soto. To the halo of legend and romance add interesting scenery of mountain, glen, stream and valley, pure  air, and

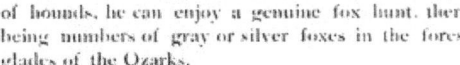

springs that emit a thermal flood of health-bearing and life-renewing waters, and you have the conditions of a model resort, not simply attractive to the invalid, but to the most fastidious and fashionable pleasure-seeker, the poet and painter.

The Indians were familiar with the curative properties of the waters before ever they were discovered by the white men, and doubtless their reputation among the former spread over the entire continent. From the accounts in history, it is reasonable to suppose that Ponce de Leon conceived his idea of a Fountain of Youth in the New World from Indian traditions of Hot Springs, which found their way from mouth to mouth to Florida, the scene of his explorations. Later, De Soto, in seeking to establish his predecessor's dream, discovered the Mississippi river, and traveled extensively in this region; and it is related that in his wanderings he came to some hot lakes where he sojourned for the winter, and in the spring, in starting out toward the southeast, died somewhere near Helena, Arkansas. There are no other hot lakes in this vicinity or anywhere in this part of the country, and we are forced to the conclusion that he spent the winter at Hot Springs, the actual Fountain of Youth he was seeking. To account for there being lakes here, it is known that the Indians, who were aware of the healing powers of the waters, came from all the neighboring tribes to bathe. Observation has disclosed that an Indian never bathes for cleanliness; therefore, it is correct to infer that he used the waters to alleviate the sufferings to which, as the son of the forest, he was natural heir. To form pools in which to plunge, he built dams across the valley, the remains of which were still visible when the government undertook the work of clearing out the valley and forcing the creek under ground.

Many Indian legends exist concerning the springs, but the most interesting and the only one which we shall give space to relate is the legend of how the water became hot.

Long ago the Kanawagas were a powerful nation. They were mighty hunters, untiring in the chase and fierce in war. They were the favored people of the Great Spirit. Their huge statures, fierce faces, sinews like hickory knots, and unequaled skill with the bow and arrow and tomahawk, which was of such great size and weight that the strongest of their enemies could scarcely raise the blade to the level of their plumes, struck terror to the hearts of their foes and made them like women in battle. By means of their superior prowess they conquered and held all of the unsurpassed hunting grounds of the mountain and valley country, from the great river, the Father of Waters, in the direction of the sunset, to the great desert plains. It was a beautiful region of mountain, lake and river, where the sun scarcely hid his face from the gaze of his children from one year's end to another, and game, fowl and fish were so plentiful that the lives of this favored people were never in danger from famine, and the warriors had time to pursue and conquer their enemies. In summer the mountains and valleys and plains were covered with wild flowers, and the ears of maize burst their husks and yielded abundant stores to the cultivation of the women of the tribe. In winter the warm valleys of the mountains afforded shelter from the storms and snows that raged when the Spirit of the Clouds was angry. Here they spread their wigwams for the short winter. This was a great people and prosperous, but a fatal day came. Prosperity and happiness were turned to lamentations and grief.

A terrible disease had fastened itself upon the members of this great and powerful tribe, and spread from one to another with fatal rapidity. Nearly all the strong men were stricken and helpless, and many were dying daily from the terrible scourge. The hunters forsook the chase, the warpath was deserted, and desolation marked the whole face of the country that was once so prosperous and smiling. The flowers came back, as if to mock at the agony of the afflicted tribe; but the corn grew not, for none had been planted. Despair superseded the last ray of hope, and a great people awaited their extermination with a stolidity becoming their character. Pure cool water only could allay their suffering in any degree, and, if as from common impulse, the survivors of the tribe dragged themselves together, the stronger assisting the weaker, to a valley of the mountains where the waters gushed forth in large quantities from numerous springs up on the mountain side, and came down in cooling, sparkling streams to the valley, making cheerful music, in striking contrast to the sad scenes below. Here the doomed tribe assembled and allayed the tortures of their fevered bodies with cooling draughts. The old and wise men of the tribe said that the Great Spirit was offended because in their prosperity they had failed to acknowledge him as the source of their greatness and power. They held dances and afflicted themselves with agonizing tortures in their efforts to appease his displeasure and restore themselves to his favor and to health again.

WHEELER'S FORD AND FERRY, ON ROAD TO BEAR MOUNTAIN.

It is supposed their beseechings were answered, for one bright afternoon, as the sun was going down again on their despair and helplessness, thin tongues of vapor were seen to issue with the water from the springs. They were too weak and indifferent to notice it at first, but the volume increased and was soon accompanied by a hissing sound, and the waters that had heretofore been cool, first became warm to the touch and afterward coursed in a heated stream down the mountain side. Some thought the last comfort to their suffering was cut off, but the wise men saw in this an end to their afflictions. The Great Spirit had breathed his healing breath into the waters, and they ordered all the people to bathe their bodies in the cleansing flood. The cure was marvelous in its rapidity, and in a few days the whole tribe was restored to its former health, and forever since the waters have poured forth a healing flood for the benefit of all mankind. They were called thereafter, No-wa-say-lon, or the Breath of Healing, by which name they were known long after the advent of the white man.

A tradition which seems to follow the foregoing legend in point of time, is to the effect that warfare and strife of any kind was never carried on here. The valley was called "Man-a-ta-ka" which signifies a "Place of Peace." It was neutral ground and here all hostilities were suspended, all met in friendship and smoked the pipe of peace. In this valley was absolute immunity from harm. It was religiously believed that if a life of any kind was taken here, even of the birds and beasts, or if strife was carried on, the Great Spirit would withdraw his breath from the waters, and they would forevermore be cold and without life and healing. Small wonder that the savage bosom was filled with sacred awe of the place, that the animals were tame and fearless, that the squirrels and smaller beasts disported unterrified on the mountain side, that the shrill war-whoop never sent terror to the breasts of defenceless women and children, that the hatchet was never raised above the head of an enemy, and the twang of the bow-string and the whiz of its deadly missile were never heard.

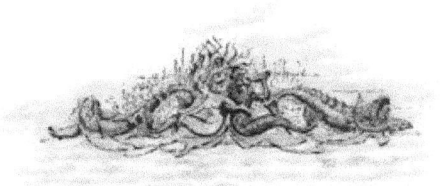

Report of the Health Department,

City of Hot Springs, Arkansas, 1892.

PER CENT OF DEATHS IN A FEW OTHER CITIES AND TOWNS.

City	Period	Rate	City	Period	Rate
Albany, N. Y.,	for first 9 months, 1891	25.00	Nashville, Tenn.,	for first 11 months, 1891	22.80
Bridgeport, Conn.,	" 11 "	19.30	New Orleans, La.,	" 11 "	26.51
Baltimore, Md.,	" 11 "	21.33	Philadelphia, Pa.,	" 11 "	20.50
Buffalo, N. Y.,	" 11 "	23.71	Paterson, N. J.,	" 11 "	21.66
Brooklyn, N. Y.,	" 11 "	24.51	Pittsburgh, Pa.,	" 9 "	25.48
Concord, N. H.,	" 9 "	17.34	Providence, R. I.,	" 8 "	19.12
Charleston, S. C.,	" 11 "	28.73	Pensacola, Fla.,	" 11 "	17.27
Cincinnati, Ohio,	" 11 "	20.88	Richmond, Va.,	" 11 "	24.65
Chattanooga, Tenn.,	" 11 "	22.72	Sacramento, Cal.,	" 8 "	17.85
Chicago, Ill.,	" 9 "	23.70	San Antonio, Tex.,	" 4 "	19.01
Detroit, Mich.,	" 11 "	19.13	St. Louis, Mo.,	" 8 "	19.80
Dayton, Ohio,	" 11 "	17.77	Troy, N. Y.,	" 4 "	27.12
Evansville, Ind.,	" 11 "	16.31	Washington, D. C.,	" 11 "	21.12
Galveston, Tex.,	" 11 "	17.25	Wilmington, Del.,	" 4 "	23.27
Hartford, Conn.,	" 11 "	21.28	Ottawa, Can.,	" 9 "	21.75
Knoxville, Tenn.,	" 11 "	16.27	Hartford,	" 11 "	23.97
Los Angeles, Cal.,	" 11 "	12.69	Hamilton,	" 9 "	14.72
Louisville, Ky.,	" 11 "	11.30	Halifax,	" 9 "	21.46
Lynchburg, Va.,	" 11 "	20.93	Montreal,	" 9 "	27.99
Milwaukee, Wis.,	" 11 "	20.57	Quebec,	" 9 "	41.72
Mobile, Ala.,	" 11 "	23.28	St. John,	" 9 "	17.80
Macon, Ga.,	" 11 "	21.35	Toronto,	" 9 "	17.34
Newport, N. J.,	" 11 "	20.30	Hot Springs, Residents		10.20
New Haven, Conn.,	" 8 "	19.12	" Visitors and Residents		12.82
New York, N. Y.,	" 11 "	27.71			

AREA OF CITY AND EXTENT OF PUBLIC IMPROVEMENTS.

Population (resident and visiting)	21,000
Residents	15,000
No. of visitors during the year	55,000
No. of Daily Papers	3
No. of Weekly Papers	4
No. of Monthlies, Illustrated	1
No. of Job Printing Offices	4
Medical Journal	1
No. of Churches	18
No. of Schools	10
No. of Hotels, Boarding Houses and Furnished Houses, more than	500
No. of Banks	3
No. of Drug Stores	22
No. of Physicians	77
No. of Planing Mills and Sash and Door Factories	3
No. of Grist Mills	1
No. of Miles in City	5
No. of Acres in City	3,169
No. of Acres in Parks, including Government Reservation	888.7
Lineal Miles of Street	70
No. of Miles of Street Railway	10
No. of Miles of Main Sewer	6.13
Capacity Arctic Ice Factory per day in Tons	28
" Cold Storage " "	15
" Valley " "	15
No. of Miles of Gas Main	4½
No. of Miles of Water Main	13
No. of Fire Hydrants	73
Capacity of Water Works per day in gallons	2,250,000
No. of Miles of Telephone Wire	100
No. of Miles of Electric Light Wire	22
No. of Hot Springs	72
No. of Bath Houses	20
No. of Steam Laundries	2

※ REMARKS. ※

RAINFALL.

	1888	1889	1890	1891
No. of inches	63.50	50.77	79.93	70.86

1888	Lowest temperature	13° above zero.	
"	Highest	"	96°
"	Mean	"	55.17°
"	Range of	"	83°
1889	Lowest	"	14°
"	Highest	"	89°
"	Mean	"	68.08°
1890	Lowest	"	13°
"	Highest	"	80°
"	Range of	"	67°
1891	Lowest	"	16°
"	Highest	"	98°
"	Range of	"	84°

ELEVATION.

In Valley, No. feet above Gulf of Mexico	608.5
Top of mountain on either side	1,200
Latitude	34° 31′ N
Longitude	92° 50′ W

AVERAGE AGE AT DEATH.

1889	37 years, 8 months and 6 days.
1890	34 years, 3 months and 4 days.
1891	33 years, 6 months and 4 days.

PER CENT OF DEATHS.

	1888	1889	1890	1891
Visitors and Residents	8.10	10.25	9.35	12.82
Residents	5.92	7.64	7.12	10.20

www.ingramcontent.com/pod-product-compliance
Lightning Source LLC
Chambersburg PA
CBHW031608110426
42742CB00037B/1332